SQL SERVER INTERVIEW QUESTIONS & ANSWERS

Written By Neelan Joachimpillai

NOTE TO READERS

The questions within this book have been compiled based on theoretical and real life interviews from several sources. By utilizing this book you should be able to identify key areas you need to review. I personally use these questions as a refresher for a lot of things I have learnt in regards to the SQL Server suite. In this release I improved the layout and fixed a few wording mistakes. This book is tailored to SQL Server 2008 R2, although most questions are generic and can be used to answer questions across versions.

Email suggestions and questions with regards to this book to:

books@otbxsolutions.ca

Regards,
Neelan Joachimpillai

OLTP Data Modelling Questions

1. What is an ER Diagram?
An ER diagram or Entity-Relationship diagram is a special picture used to represent the requirements and assumptions in a system from a top down perspective. It shows the relations between entities (tables) in a database.

2. What is a prime attribute?
A prime attribute is a attribute that is part of a candidate key.

3. What are the properties of a transaction?
The ACID properties. Atomicity, Consistency, Isolation, and Durability.

4. What is a non-prime attribute?
A non-prime attribute is an attribute that is not a part of a candidate key.

5. What is Atomicity?
This means the transaction finish completely, or it will not occur at all.

6. What is Consistency?

Consistency means that the transaction will repeat in a predictable way each time it is performed.

7. What is Isolation?
The data the transactions are independent of each other. The success of one transaction doesn't depend on the success of another.

8. What is Durability?
Guarantees that the database will keep track of pending changes so that the server will be able to recover if an error occurs.

9. What is a DBMS?
A DBMS is a set of software programs used to manage and interact with databases.

10. What is a RDBMS?
It is a set of software programs used to interact with and manage relational databases. Relational databases are databases that contain tables.

11. What is business intelligence?
Refers to computer-based techniques used in identifying, extracting, and analyzing business data, such as sales revenue by products

4

and/or departments, or by associated costs and incomes.

12. What is normalization?
Database normalization is the process of organizing the fields and tables of a relational database to minimize redundancy and dependency.

13. What is a relationship?
The way in which two or more concepts/entities are connected, or the state of being connected.

14. What are the different types of relationships?
One to one, one to many, many to many, many to fixed cardinality.

15. What is the difference between a OLTP and database?
An OLTP is the process of gathering the data from the users, and a database is the initial information.

16. What are the different kinds of relationships?
Identifying and non-identifying.

17. What is an entity?
Something that exists by itself, although it need not be of material existence.

18. What is a conjunction table?
A table that is composed of foreign keys that points to other tables.

19. What is a relational attribute?
An attribute that would not exist if it were not for the existence of a relation.

20. What are associative entities?
An associative entity is a conceptual concept. An associative entity can be thought of as both an entity and a relationship since it encapsulates properties from both. It is a relationship since it is serving to join two or more entities together, but it is also an entity since it may have its own properties.

21. What is the difference between a derived attribute, derived persistent attribute, and computed column?
A derived attribute is a attribute that is obtained from the values of other existing columns and does not exist on it's own. A derived persistent attribute is a derived

attribute that is stored. A computed attribute is a attribute that is computed from internal system values.

22. What are the types of attributes?
Simple, composite (split into columns), multi-valued (becomes a separate table), derived, computed, derived persistent.

23. Is the relationship between a strong and weak entity always identifying?
Yes, this is the requirement.

24. Do stand alone tables have cardinality?
No.

25. What is a simple key?
It is a key that in composed of one attribute.

26. Give/ recite the types of UDF functions.
Scalar, In-line, Multi

SQL Questions

27. Describe what you know about PK, FK, and UK.

Primary keys – Unique clustered index by default, doesn't accept null values, only one primary key per table.
Foreign Key – References a primary key column. Can have null values. Enforces referential integrity.
Unique key – Can have more than one per table. Can have null values. Cannot have repeating values. Maximum of 999 clustered indexes per table.

28. What do you mean by CTEs? How will you use it?
CTEs also known as common table expressions are used to create a temporary table that will only exist for the duration of a query. They are used to create a temporary table whose content you can reference in order to simplify a queries structure.

29. What is a sparse column?
It is a column that is optimized for holding null values.

30. What would the command: DENY CREATE TABLE TO Peter do?
It wouldn't allow the user Peter to perform the operation CREATE TABLE regardless of his role.

31. What does the command: GRANT SELECT ON project TO Peter do?

It will allow the SELECT operation on the table 'project' by Peter.

32. What does the command: REVOKE GRANT SELECT ON project TO Peter do?

It will revoke the permission granted on that table to Peter.

33. New commands in SQL 2008?

Database encryption, CDCs tables – For on the fly auditing of tables, Merge operation, INSERT INTO – To bulk insert into a table from another table, Hierarchy attributes, Filter indexes, C like operations for numbers, resource management, Intellisense – For making programming easier in SSMS, Execution Plan Freezing – To freeze in place how a query is executed.

34. What is new in SQL 2008 R2?

PowerPivot, maps, sparklines, data bars, and indicators to depict data.

35. What is faster? A table variable or temporary table?

A table variable is faster in most cases since it is held in memory while a temporary table is stored on disk. However, when the table variable's size exceeds memory size the two table types tend to perform similarly.

36. How big is a tinyint, smallint, int, and bigint?
1 byte, 2 bytes, 4 bytes, and 8 bytes.

37. What does @@trancount do?
It will give you the number of active transactions for the current user.

38. What are the drawbacks of CTEs?
It is query bound …

39. What is the transaction log?
It keeps a record of all activities that occur during a transaction, and is used to roll back changes.

40. What are before images, after images, undo activities and redo activities in relation to transactions?
Before images refers to the changes that are rolled back on if a transaction is rolled back. After images are used to roll forward and

enforce a transaction. Using the before images is called the undo activity. Using after images is called the redo activity.

41. What are shared, exclusive and update locks?

A shared lock, locks a row so that it can only be read. An exclusive lock locks a row so that only one operation can be performed on it at a time. An update lock basically has the ability to convert a shared lock into an exclusive lock.

42. What does WITH TIES do?

If you use TOP 3 WITH TIES *, it will return the rows, that have a similarity in each of their columns with any of the column values from the returned result set.

43. How can you get a deadlock in SQL?

By concurrently running the same resources that access the same information in a transaction.

44. What is LOCK_TIMEOUT used for?

It is used for determining the amount of time that the system will wait for a lock to be released.

45. What is the ANY predicate used for?

```
SELECT * FROM emp_table WHERE
enter_date > ANY (SELECT enter_date FROM
works_on)
```

46. What is the ALL predicate used for?

```
SELECT * FROM emp_table WHERE
enter_date > ALL (SELECT enter_date FROM
works_on)
```

47. What are some control flow statements in SQL?

while, if, case, for each etc..

48. What is the EXISTS function used for?

It is used to determine whether a query returns one or more rows. If it does, the EXIST function returns TRUE, otherwise, it will return FALSE.

49. State the different types of procedures.

System, User, CLR, Extended

50. How can you get XML results in SQL server?

You can use the FOR XML command for SQL to return XML results rather than the regular row based results. You have to specify the types (ROW, AUTO, EXPLICIT).

51. What table holds the fragmentation statistics of the database?

sys.dm_db_index_physical_stats

52. What is a Filtered Index?

A filtered index is a non-clustered index that creates a B-tree based on a particular predicate.

53. Is truncate in the log file?

Yes. However, it doesn't provide any more information about the rows that were removed. Therefore, it only tells you that the table was truncated.

54. What are the properties of a transaction?

(ACID) Atomicity, Consistency, Isolation and Durability.

55. What is a stored procedure?

It's nothing but a set of T-SQL statements combined into a single batch command. In MSSQL it has a precompiled query plan and thus tends to perform more efficiently than if you were not to use a stored procedure. It is basically like a Macro so when you invoke the Stored procedure, you actually run a set of statements.

56. What is classifying?
Giving the data attributes.

57. What is recording?
Making copies of the data.

58. What is data processing?
Operations on the data.

59. What is data storing?
How you store the data.

60. How do you add a linked server?
You can either use a stored procedure or you can right click on server objects and click 'add linked server'.

61. What are the data processing modes?
Online, batch, and real time.

62. What is the syntax of a stored procedure?
CREATE PROCEDURE OrderSummary
@MaxQuantity INT OUTPUT AS
SELECT Ord.EmployeeID, SummSales =
SUM(OrDet.UnitPrice * OrDet.Quantity)
FROM Orders AS Ord

```
    JOIN [Order Details] AS OrDet ON
(Ord.OrderID = OrDet.OrderID)
GROUP BY Ord.EmployeeID
ORDER BY Ord.EmployeeID
```

63. Write an inline UDF functions.

```
CREATE FUNCTION udf_inline
RETURN TABLE
  AS ( SELECT * FROM
HumanResources.Employee )
```

64. Are triggers automatic or not automatic? What is the syntax for creating a trigger?

Triggers are run every time a specific operation is performed on a table.

The syntax for creating a trigger is as follows:

```
CREATE TRIGGER trg_t1 ON test_trigger
AFTER INSERT, DELETE
  AS
  BEGIN
    PRINT 'Inside after trigger ...'
  END
```

65. What is a view? Describe it's advantages and disadvantages.

A view is a virtual table based on the result-set of an SQL statement.

Advantages

- It provides more security
- It can protect the underlying table from alterations.
- It can hide columns and restrict the perspective.
- Reduces unnecessary messy code
Disadvantages
- It is slower since you are not querying the table directly
- It is an object so it takes up space
- You cannot use DML operations on it

66. What are the steps in creating an indexed view?

The ANSI_NULLS option must have been set to ON for the execution of all CREATE TABLE statements that create tables referenced by the view.

The view must not reference any other views, only base tables.

All base tables referenced by the view must be in the same database as the view and have the same owner as the view.

The view must be created with the SCHEMABINDING option. Schema binding binds the view to the schema of the underlying base tables.

User-defined functions referenced in the view must have been created with the SCHEMABINDING option.
Tables and user-defined functions must be referenced by two-part names in the view. One-part, three-part, and four-part names are not allowed.
All functions referenced by expressions in the view must be deterministic. The IsDeterministic property of the OBJECTPROPERTY function reports whether a user-defined function is deterministic.

67. What is the difference between a unique clustered index and clustered index?

The clustered allows duplicates while the unique clustered index does not. Clustered indexes do not uniquely sort the data.

68. What is the difference between an inline function and multiline function?

Multi-statement UDFs let you include multiple statements in the UDF. The UDF returns a result set that's populated inside the function, as I discuss later. Inline table-valued UDFs can include only one SELECT statement. Multi-statement table-valued UDFs can add flexibility to many T-SQL solutions, but inline

UDFs are almost always more efficient than comparable multi-statement UDF solutions.

69. How did you use a CTE in your last job?
CTEs are good because you can declare a CTE once rather than having to declare it more than once. Furthermore, you can reference the CTE within itself as well as have more than one CTE for a particular transaction. This offers an advantage over sub-queries because when you use a sub-query, you have to declare it whenever you have to use it.

71. Write the syntax for a transaction.
BEGIN TRAN tran1
COMMIT TRAN tran1
SAVE TRAN tran1
ROLLBACK TRAN tran1

72. What are the types of transactions?
Commit, Save, Roll Back, Begin

73. What is OPENQUERY in T-SQL?
OPENQUERY is an operation that is used in order to query linked server databases (remote databases).

74. What is the difference between mean, median and mode?

The mean is the arithmetic average of a set of values. The median, is the value separating the higher values from the lower values. The mode is the value that occurs the most commonly.

75. Why do we have transactions?
To correct recovery data and guarantee that the data will be consistent.

76. How do design a database?
   - Identify the business transactions to be tracked.
      - Identify the entities and attributes.
      - Identify the relationships
      - Normalize.
      - Convert logical into physical

77. What is the difference between a function and a view?
A function can have DML and DDL operations while a view is simply a result sets query.

78. What is the difference between a stored procedure and user defined functions?
   Difference between SPs and UDFs:
   1) SP executed using EXEC while you CALL UDF (you can use SELECT)

2) UDF must return something while SPs don't have to.

3) SP can have output params, while UDFs cannot have output params. Both can have returns

though.

4) Returns in SP are scalar, in UDF, outputs are 2 dimensional.

5) In UDFs you cannot have TRY/CATCH statements.

79. In what kind of situations can you use user defined functions?
When you want to utilize the table output.

80. What does WAIT FOR DELAY '00:00:02' do?
It introduces a manual delay into the system before the next line in the query is processed. This allows for less network congestion and resource conflicts.

81. What types of DDL triggers are there?
Only the AFTER trigger.

82. What are the RANKING functions?
Rank, Dense Rank, ntile, rownumber

83. What does rank do?

It will give a unique rank based on the value of a column. It will skip if you have repeating values.

84. Rownumber?
It will give a unique value to each row.

85. Denserank?
Denserank will give a unique rank based on the value of a column, however, it will continue at the consecutive numbers after repeats.

86. NTile?
NTile will partition the values into equivalent parts.

87. What is the difference between a Union and Union all in SQL?
Union specifies that multiple result sets are to be combined and returned as a single result set.

Which means it would only show one row of duplicate rows. Union all incorporates all rows into

the result. This includes duplicates. Which means it would keep the duplicate rows, keep all rows.

88. What is the difference between a non-cluster and clustered index?

A clustered index is ordered and in it's leaf nodes contains data. A table can only have one clustered index at a time. Non-clustered indexes have pointers to a heap in their leaf nodes and you can have more than one non-clustered index in a table.

89. What are cursors?

A cursor is a database object used to iterate through a result set one row at a time.

90. Which TCP/IP port does SQL Server run on?

1433

91. Can we use Truncate command on a table which is referenced by FOREIGN KEY?

No.

92. What is the use of DBCC commands?

DBCC commands stands for database console commands. DBCC commands take input parameters and return values. All DBCC command parameters can accept both Unicode and DBCS literals.

93. Can you give me some DBCC command options?(Database consistency check)

DBCC CHECKALLOC
DBCC CHECKCATALOG
DBCC CHECKTABLE
DBCC CHECKDB
DBCC CHECKFILEGROUP

94. What command do we use to rename a db?

sp_renamedb. This command will be phased out in upcoming releases of MS SQL.

95. What is the difference between a HAVING CLAUSE and a WHERE CLAUSE?

HAVING is used for aggregate functions, while the WHERE clause is used for non-aggregate queries.

96. What do you mean by COLLATION?

A SQL Server collation defines how the database engine stores and operates on character and Unicode data.

97. What is a Join in SQL Server?

By using joins, you can retrieve data from two or more tables based on logical relationships between the tables. Joins indicate how

Microsoft SQL Server should use data from one table to select the rows in another table.

98. What is a left join in SQL Server?
A left join will return the data set from the left table regardless of whether there is a match in the right table.

99. Can you explain the types of Joins that we can have with SQL Server?
Inner join, Full outer join, left outer join, right outer join, and cross join.

100. When do you use SQL Profiler?
Microsoft SQL Server Profiler is a graphical user interface to SQL Trace for monitoring an instance of the Database Engine or Analysis Services. You can capture and save data about each event to a file or table to analyze later. For example, you can monitor a production environment to see which stored procedures are affecting performance by executing too slowly.

101. What is a Linked Server?
A linked server configuration enables SQL Server to execute commands against OLE DB data source on remote servers.

102. When defining a FORIEGN KEY, what does the constraint CASCADE DELETE DO?

If the primary key is deleted or changed so will be the records referencing it.

102. What is the distributed transaction coordinator?

The Distributed Transaction Coordinator (MSDTC) service is a component of modern versions of Microsoft Windows that is responsible for coordinating transactions that span multiple resource managers, such as databases, message queues, and file systems.

103. Can you link only other SQL Servers or any database servers such as Oracle?

As long as you have the appropriate ODBC connection to the database, you will be able to connect to it.

104. What are the authentication modes in SQL Server?

1. Windows Authentication mode.2. Windows and SQL Authentication mode

105. Where do you think the users names and passwords will be stored in SQL server?

In the system database master.

106. What is log shipping?

Log shipping allows you to automatically send transaction log backups from a primary database on a primary server instance to one or more secondary databases on separate secondary server instances. The transaction log backups are applied to each of the secondary databases individually. An optional third server instance, known as the monitor server, records the history and status of backup and restore operations and, optionally, raises alerts if these operations fail to occur as scheduled.

107. What is BCP? When do we use it?

The bcp utility bulk copies data between an instance of Microsoft SQL Server and a data file in a user-specified format. The bcp utility can be used to import large numbers of new rows into SQL Server tables or to export data out of tables into data files. Except when used with the query out option, the utility requires no knowledge of Transact-SQL. To import data into a table, you must either use a format file created for that table or understand the structure of the table and the types of data that are valid for its columns.

108. What should we do to copy the tables, schema and views from one SQL Server to another?

You can either use SQL's export/import wizard, or you can generate an export script and run it against the other server.

109. What are the four main query statements?

SELECT, UPDATE, INSERT, and DELETE

110. What is a sub-query? When would you use one?

It is a query within another query. It is used when you know how to search for a value using a SELECT statement, but do not know the exact value.

111. What is the Full-Text Search Engine used for?

It is a engine in SQL Server used for text searching columns in order to speed up queries. It has two primary functions:

CONTAINS and FREETEXT. Furthermore, it has a dictionary search in order to find abstract word relations.

111. What is a correlated sub-query?

A correlated sub-query is a query that has an inner sub-query that is dependent on the results of
the outer sub-query, and a outer sub-query that is dependent on the results of inner sub-query.

112. What is READUNCOMMITTED?

Implements dirty read, or isolation level 0 locking, which means that no shared locks are issued
and no exclusive locks are honored. When this option is set, it is possible to read uncommitted or
dirty data; values in the data can be changed and rows can appear or disappear in the data set
before the end of the transaction.

113. What are three SQL keywords used to change or set someone's permissions?

GRANT, DENY, and REVOKE

114. What is referential integrity? What are the advantages of it?

For referential integrity to hold in a relational database, any field in a table that is declared a foreign key can contain only values from a parent table's primary key or a candidate key.

115. What is database normalization?

Database normalization is a design methodology used to reduce redundancy and INSERT, DELETE and UPDATE errors when dealing with databases. It makes operations on the database more memory efficient and finds a balance between speed and memory use.

116. What is SQL server agent?

SQL Server Agent is a Microsoft Windows service that executes scheduled administrative tasks, which are called jobs. SQL Server Agent uses SQL Server to store job information. Jobs contain one or more job steps. Each step contains its own task, for example, backing up a database. SQL Server Agent can run a job on a schedule, in response to a specific event, or on demand.

117. What is the STUFF function and how does it differ from the REPLACE function?

Both STUFF and REPLACE are used to replace characters in a string.

SELECT REPLACE('abcdef','ab','xx')
results in xxcdef

SELECT REPLACE('defdefdef','def','abc')
results in abcabcabc

We cannot replace a specific occurrence of "def" using REPLACE.

SELECT STUFF('defdefdef',4, 3,'abc')
results in defabcdef

118. What does it mean to have quoted_identifier on? What are the implications of having it off?

When SET QUOTED_IDENTIFIER is ON (default), identifiers can be delimited by double quotation marks, and literals must be delimited by single quotation marks. When SET QUOTED_IDENTIFIER is OFF, identifiers cannot be quoted and must follow all Transact-SQL rules for identifiers.

119. What are the different types of replication? How are they used?

Replication is a set of technologies for copying and distributing data and database objects from one database to another and then synchronizing between databases to maintain consistency. Using replication, you can distribute data to different locations and to remote or mobile users over local and wide area networks, dial-up connections, wireless connections, and the Internet.

Transactional replication is typically used in server-to-server scenarios that require high throughput, including: improving scalability and availability; data warehousing and reporting; integrating data from multiple sites; integrating heterogeneous data; and offloading batch processing. Merge replication is primarily designed for mobile applications or distributed server
applications that have possible data conflicts. Common scenarios include: exchanging data with
mobile users; consumer point of sale applications; and integration of data from multiple sites.

Snapshot replication is used to provide the initial data set for transactional and merge replication;
it can also be used when complete refreshes of data are appropriate.

120. What is the difference between a local and a global variable?

Local variables are user defined variables that exist for the duration of a query. Global variables are
internal variables in SQL Server that contain system and operation related values.

121. What is the difference between a local temporary table and a global temporary table?

A local temporary table exists only for the duration of a connection or, if defined inside a compound statement, for the duration of the compound statement. A global temporary table remains in the database permanently, but the rows exist only within a given connection. When connection are closed, the data in the global temporary table disappears. However, the table definition remains with the database for access when database is opened next time.

122. What are cursors? Name four types of cursors and when each one would be applied?

   a. Base table
   b. Static
   c. Forward-only

d. Forward-only/Read-only

e. Keyset-driven

123. What is the purpose of UPDATE STATISTICS?

Update statistics is use to generate background statistics on a table or database. The statistics are used by query optimizer in order to speed up the performance of queries.

124. How do you use DBCC statements to monitor various aspects of a SQL server installation?

Database Console Commands for SQL Server. They allow you to check the physical integrity of a database.

125. How do you load large data to the SQL server database?

You can use a BULK INSERT.

126. How do you check the performance of a query and how do you optimize it?

You can use SET STATISTICS TIME ON or SQL Profiler to accomplish it. You can optimize the queries by using the Database Tuning Advisor suggestions or Hints.

127. What is @@rowcount?

Returns the number of rows affected by the last statement. If the number of rows is more than 2 billion, use ROWCOUNT_BIG.

128. What is the difference between a function and stored procedure in general?

Difference between SPs and UDFs:

1) SP executed using EXEC while you CALL UDF (you dont EXEC, use SELECT)

2) UDF must return something while SPs don't have to.

3) SP can have output params, while UDFs cannot have output params. Both can have returns
though.

4) Returns in SP are scalar, in UDF, outputs are 2 dimensional.

5) In UDFs you cannot have TRY/CATCH statements.

6) UDFs can return tables.

129. What is a spatial index?

It is arranged in an R-tree structure. It is used to categorize 2D date. It is used mostly for geometry and coordinates. It was introduced in SQL 2008.

130. When should you backup a production database?

You should back it up when you first create it, when you creates indexes, when you clear transactions, and when you perform non-logged operations.

131. What is @@NESTLEVEL?
Returns the nesting level of the current stored procedure execution (initially 0) on the local server.

132. Difference between Cluster and Non-Cluster index?
A clustered index organizes data in the leaf nodes of it's B-tree and you can only have one on any given table. A non-clustered index has pointers to the heap in it's leaf nodes.

133. What is @@SERVERNAME?
Returns the name of the local server that is running SQL Server.

134. What is data integrity?
Data integrity is an important feature in SQL Server. When used properly, it ensures that data is accurate, correct, and valid. It also acts as a trap for otherwise undetectable bugs within applications.

135. Name the types of constraints.

Primary Key, Foreign Key, Unique, NOT NULL, CHECK, DEFAULT, Data Type (According To Ali)

136. What is the basic functions for master, model, msdb, tempdb databases?

The Master database holds information for all databases located on the SQL Server instance and is the glue that holds the engine together. Because SQL Server cannot start without a functioning master database, you must administer this database with care.

The msdb database stores information regarding database backups, SQL Agent information, DTS packages, SQL Server jobs, and some replication information such as for log shipping.

The tempdb holds temporary objects such as global and local temporary tables and stored procedures.

The model is essentially a template database used in the creation of any new user database created in the instance.

137. How would you Update the rows which are divisible by 10, given a set of numbers in column?

Using the modulus operator to find which numbers are divisible by 10.

138. What is the difference between a primary and foreign key?

Primary keys are the unique identifiers for each row. They must contain unique values and cannot be null. Due to their importance in relational databases, Primary keys are the most fundamental of all keys and constraints. A table can have only one Primary key. Foreign keys are both a method of ensuring data integrity and a manifestation of the relationship between tables.

139. What do the 1NF, 2NF and 3NF forms of normalization entail?

The 1NF form gets rid of redundant records and repeating attributes.

The 2NF form gets rid of partial dependencies.

The 3NF form gets rid of transient dependencies.

140. What is a bulk insert?

It is when you take a flat file and dump it into a SQL server table.

141. What are the DML operations?

Operations like SELECT, DELETE, UPDATE, INSERT.

142. What are the DDL operations?
Schema level. CREATE, ALTER, DROP, TRUNCATE.

143. What is a partition key?
It is a column on which the partitioning of a table is based.

144. What is a partition scheme?
They are used to map partitioned row indexes to partition files based on a function.

145. What is a partition function?
Partitions a table based on the values of it's partition index or key.

146. How will you achieve particular goals with partitioning?
If you want to improve performance for particular queries, distribute the rows evenly among partitions. Make sure each partition is stored in a different file group in different disk devices. Partition based on a column that doesn't frequently change values.

147. What is collocation?
It is when you partition two tables using the same partition function.

148. What do you have to be aware of when performing a bulk insert?

You have to make sure that the input in the file is clean since there is no validation or error logging. Furthermore, you must assure that the table being dumped into has the same number of columns and schema as the flat file.

149. When utilizing the default constraint, if you do not enter a value, will the column be populated with the default constraint specified?

No. The default constraint requires you enter a specific value.

150. How many SQL queries can be nested?

32

151. Why can we only have a single clustered index?

Because the data in the table can only be sorted in one way. It is physical be sorted and stored.

152. What is a faster way to implement an OR constraint in the same variable?

By utilizing the WHERE IN argument in your SELECT statement.

153. What is a direct configuration?
It is when you directly give the configuration type and path to where the properties are stored.

154. What is the problem with direct configuration files?
When you move the configuration file from one computer to another, the path may change. There is no guarantee that the configuration string will be the same.

155. What is sqlcmd?
Sqlcmd is a command prompt utility used to run SQL commands from the command line.

156. What is a dataflow vs control flow?
A dataflow contains all the tasks that perform operations on the data, while a control flow shows the flow of data through the system.

157. How can you mimic a materialized view in MS SQL?
By using indexed views.

158. How do you rebuild an index?

By using the ALTER TABLE command with a REBUILD INDEX post fix. Below is an example:

```
ALTER INDEX
PK_ProductPhoto_ProductPhotoID
  ON Production.ProductPhoto
  REORGANIZE/REBUILD
```

OLAP Data Modelling Questions

159. What is a dimension table?
A dimension table is a table that contains the attributes of the facts or measures from a fact table.

160. How would you handle late arriving data?
Hold on to the fact, insert a dummy row, unknown dimension, inferred dimension (you need
enough information to make a natural key). Reference: http://blog.todmeansfox.com/2008/07/10/etl-subsystem-16-late-arriving-data-handler/

161. What is dimensional modelling?

The act of designing a data warehouse. It is different from the ER modelling used with databases. It
is composed of four steps:
Identify the business processes to be analyzed.
Identify the dimensions and their attributes.
Identify the grain.
Identify the facts.

162. How do you choose your business processes?
You have to make sure the business processes make money if they are analyzed. Things that have worth.

163. How long does it usually take to create a data mart?
Three to five months. Cost is about $200,000.

164. How long does it take to create a data warehouse?
Two years. Cost is about $5 million.

165. What is the difference between a dimensional modelling and relational modelling?

Dimensional modelling is used to design data warehouses, while relational modelling is used to create OLTPs.

166. How do you choose your measures?
The measures are usually quantity/amount or cost. They are ALWAYS a numeric value.

167. How do you choose your granularity?
Most atomic property in your OLTP.
Something you cannot break down any further for a fact.

168. What is a data warehouse?
A data warehouse is a repository of historical data that contains all the business facts to be analyzed.

169. What is a data mart?
A data mart is a subset of a data warehouse. It is specific to a few business processes, but not all of them.

170. What is a fact table?
The fact table contains business facts or measures and foreign keys which refer to candidate keys (normally primary keys) in the dimension tables.

171. Are fact tables normalized or not normalized?

They are inherently normalized due to their inherent properties.

172. What are the fundamental stages of data warehousing?

173. What is real time data modelling?

It is a data warehouse that is updated immediately, rather than during specific times.

174. What is a conformed dimension?

It is a dimension that is referred to by more than one fact table.

175. What is a factless fact table?

It is a conjunction table between a fact table and dimension table that has a many to many relationship.

176. What is a degenerate dimension table?

It is a dimension table that is also partly a fact table.

177. What is a junk dimension?

It is when small unrelated dimensions are lumped together when they do not belong anywhere else. This can minimize the creation of extra dimensions that would complicate the design.

178. What are the different methods of loading data warehouses?

SSIS or a stored procedure.

179. What is the difference between a view and a materialized view?

Each time you use a view, you rerun the query, while when you utilize a materialized view. It has the results separately and is only refreshed when you specify. This makes materialized views faster. However, since views are real time, they may at times hold more accurate data.

180. What is the difference between OLTP and OLAP?

OLTP is the process used to interact with the end user and gather data. OLAP is the process of analysing the large amounts of data.

181. Steps in data warehouse design?

 - Identify the business transactions to be analyzed.
   - Identify the dimensions and their attributes.
   - Determine the level of granularity.

182. What is a dimension table and how is it chosen?
Analyze the business process by breaking it down into how you would achieve satisfying the fact table.

183. What is a fact table?
A fact table is a business process that is analyzed that can earn money or produce a consumable product that can be sold.

184. What is the difference between a data warehouse and a OLAP?
A data warehouse is where all the historical data is stored. OLAP is the PROCESS of analysing that data.

185. What is a BUS matrix?
You identify common dimensions across all dimensions and combine them. It is a tool used for planning out a data warehouse according to Ralph Kimball.

186. What is the star model?
The star model is a method for developing data marts created by Ralph Kimball that only has dimensional tables connected directly to a fact table. It is optimized for speed, however, it

tends to take up more space than Inmon's snow flake model.

187. What is the snow flake model?
The snow flake model is a method for designing data marts created by Bill Inmon. It has dimensions connected to fact tables, and dimensions connected to other dimensions. It tends to be smaller than Kimball's star schema, however, it also tends to be slower.

SSIS Questions

1. What is a ETL strategy document?
It is a document that maps out the transformations, where you will be getting data from, and the transformations that need to be made in order to make your data available to the data warehouse.

2. How did you make sure that OLAP is loaded with the right data?
QA testing of the ETL processes. Test queries.

3. What is the lifespan of a checkpoint file in SSIS?
The checkpoint file will disappear after it is run and the error point has been passed.

4. Do you have row redirect in destination adapter?
Yes

5. Name the transformations that do not have error outputs?
Multicast, Union All

6. Package configuration options in SSIS?
XML, Windows Event log, Parent Package variable, SQL Server, Environmental

7. What is the difference between an environmental variable configuration and XML?
One variable held vs many variables.

8. Name a few tasks you commonly use for SSIS in the data flow.
Slowly changing dimensions, data conversion, sort, multi-cast, merge, merge join, cache transform, lookup, script component, conditional split, flat file source, OLE DB source, OLE DB command, OLE DB destination and SQL Server destination.

9. Name a few components you commonly use in the control flow.

Sequence container, group container, data flow task, file system task, for each loop container, execute package, execute process, script task, EXECUTE SQL task, analysis services processing task, and analysis services execute task.

10. What are the different log providers in SSIS?
Windows Event log, XML log, SQL Server Trace file, SQL Server Table.

11. What are the two main natures of each transformation?
Blocking (non-block, partly block, full block) and communication (async/sync)

12. Is the 'execute SEQUEL command' blocking?
It is not blocking, because it executes the command on a row by row basis. It can also at times be partially blocking.

13. What is an execution tree?
An execution tree breaks up an SSIS package into threads and dictates the order in which the threads will be executed.

14. When will you use transactions and check points in SSIS?

The transactions in T-SQL in order to roll back. Check points allow you to make a check point and throw an error when a problem occurs. If the roll back or execution of a package or container or task requires a long time, use a check point over a transaction.

15. What is the difference between a group container and a sequence container?

Group containers are just for viewing and do not have properties while a sequence container allows for you set properties on it and breaks up tasks into different task units.

16. What is the difference between a file system task and ftp task?

A file system task will allow for SSIS to manage files and folders on the local system, while a ftp task will allow for files and folders to be managed on remote ftp servers.

17. What are the different types of dimensional tables?

Role-Playing, Parent-Child, Conformed, Junk, Degenerated.

18. What is a conditional split transformation?

The conditional split splits the results of a query based on a case statement. It will split the rows based on the contents of their data.

19. What are the different transaction options?

There are supported, non-supported, and required transactions.

20. XML and SQL server configuration?

XML configurations store the SSIS package properties and variables in a XML file. SQL server configurations store SSIS package properties and variables in a SQL table.

21. Why do we have to configure an SSIS package?

It is used when you migrate a package from one system to another. It will allow for the new environment to have the same settings and variables as the old one.

22. What is in a configuration file in general?

It stores information about variables and system configuration information.

23. What is the difference between a control flow and data flow?

A dataflow is a subset of a control flow in an SSAS. Control flows dictate the order in which control tasks are performed while a dataflow dictates the flow of data within the individual control flow tasks. In the control flow you can perform dataflow tasks in parallel.

24. What are the different ways you can do error handling in SSIS?

Event handlers, fail pipeline, transactions, and check points.

25. What are the benefits of using a checkpoint in a package?

If you have an error, it will only restart from the point of error. Thus, for big operations, you will not have to repeat all the work that had been completed. Checkpoint files are XML files.

26. How many Control flows do you have in a SSIS package?

One

27. How do you pass a variable from one package to another in SSIS?

Utilizing parent package variable option in Package configurations in SSIS. Remember to set it on the child. Then from the parent, you execute the child.

28. What is a task host container in SSIS?
A task host container is an invisible container that encapsulates every task.

29. How many executables are there in SSIS?
At the top level you have the package, inside the package you have the container, and inside the container you have the task.

30. How many types of variables does SSIS have?
System variables and user variables.

31. What is the control object 'file system task' in SSIS?
It is used to perform file and directory based operations.

32. What is delay validation in SSIS?
It is set on an executable and this property will delay the validation of the task or container to runtime.

33. What is the purpose of the sequence container?

It allows you to manage a large number of tasks. Thus, any property you place on the task will apply to all of the contained elements. It allows for better and easier management of tasks.

34. What is the difference between a group and a container?

A group is for visual grouping of elements, however, it is not an executable and does not have properties. A container is an executable and has properties.

35. What is the purpose of the Execute Process Transaction?

It is for calling batch operations outside of SQL (like batch files)

36. When was SSIS created? What existed before SSIS?

SSIS was created in 2005. Before SSIS, DTS was commonly used before this.

37. What is the send mail task?

It allows you to send email as part of your SSIS task.

38. What is the script task?

This allows you to write your own code for a particular task in .NET.

39. How do you configure a script task and refer to an external variable? Remember SSIS is case sensitive.

```
Public Sub Main()
        If
File.Exists(Dts.Variables("User::path").Value)
= True Then
            Dts.Variables("User::fileExists").Value
= True
        Else
            Dts.Variables("User::fileExists").Value
= False
End If
Dts.TaskResult = ScriptResults.Success
End Sub
```

40. What is an event handler in SSIS? What is the use of an event handler? What are the events attached to?
An event handler is like an SQL trigger in SSIS. The events are attached to the SSIS executables. On a single executable, you can have as many events as there are events that exist.

41. What are the different types of log provider types in SSIS?

They are the exact same types of information stored in different formats. You can have the following provider types: Windows Event log file, text files, XML files, SQL server databases and trace files.

42. Are checkpoints only written when the package fails?

Yes.

43. What are transactions in SSIS?

SSIS uses transactions to bind the database actions that tasks perform into atomic units, and by
doing this maintain data integrity.

44. What are SSIS configuration files used for?

It is used in order to move the system configuration in between Dev and Prod environments.

45. What happens when you change SSIS's configuration file?

It will default and use the configuration file properties instead of the initial properties. The

SSIS package will work as long as there exists an identical table in the changed database.

46. What are the most common SSIS configuration types used?
Normally people only use SQL Server and XML configurations because they store more than one value and are normally used in the industry. The environmental and registry configuration files only store a single value and thus are rarely used.

47. How many configurations can you have on a package?
You can have as many configurations on a package as you want.

48. What is a configuration filter?
It is an option under the SQL configuration type. It is used to identify which property belongs to an particular package when using a single table to record the configurations settings of multiple packages.

49. What is the difference between pipelines and precedence constraints in SSIS?
Pipelines are for data flow, and precedence constraints are for the control flow.

50. What is a environment variable and how does it apply to SSIS configuration files?

The environment variable is a variable on the OS level. It is used when with SSIS configuration, you can do INDIRECT CONFIGURATION, and the variables value will be stored in the package, and all you have to do is make sure that all computers this is being deployed to will have to have the variable on them.

51. What are the steps in making an indirect configuration?

You will have to first make a direct variable by hard coding a direct variable. Then you will have to manually make an environmental variable utilizing System properties with the path of the file. Restart Visual Studios. Then reconfigure the configuration to use the environmental value by selecting the option.

52. What is an indirect configuration file?

It points to the environmental variable that holds the configuration file path, rather than directly pointing to the file in the configuration file's settings.

53. What is the task to call an SSIS package from inside another SSIS package?

By using the Execute Package Task.

54. What is variable mapping in SSIS and how is it used?

Variable mapping is used to map a general value depicted by ? to an SSIS variable in an SQL statement or Execute command.

55. What is the Parent package variable in Package Configuration used for?

It is used to pass variable values from parents to child packages. Remember that the settings have to be set on the child package.
Remember that you can only passes scalar value via SSIS.

56. Is SSIS case sensitive?

Yes

57. What is DTutil?

The DTutil command prompt utility is used to manage SQL Server Integration Services packages. The utility can copy, move, delete, or verify the existence of a package. These actions can be performed on any SSIS package that is stored in one of three locations: a Microsoft SQL Server database, the SSIS Package Store, and the file system. The storage type of the package is identified by the /SQL, /FILE, and /DTS options.

58. How do you deploy a package using SSIS?

Go to the file menu. Click on 'save a copy of <project_name> as' and fill out the required information while selecting SQL server as the package location. Select the location you want the data at and save.

59. Give an example of package deployment using the command line.

(Remember the double quotes in the path name"
dtutil /FILE
"C:\Users\ConfusedInAManlyWay\Documents\
Visual Studio
2008\Projects\March13\March13\child.dtsx"
/DestServer localhost /COPY
SQL;childpackage

60. What is DTexec.exe used for?

This is used to execute packages from the command line.

61. What are some ways to speed up SSIS?

By using SQL queries, minimizing blocking transformations, and minimizing the number of transformations.

62. What is the MAXCONCURRENTEXECUTABLES Property?

It states the maximum executables that can run on a system at a time.

63. What are the different types of transactions in SSIS and what do they do?

Supported indicates that the container does not start a transaction, but joins any transaction started by its parent container. Required indicates that the container starts a transaction. Supported indicates that the container does not start a transaction, but joins any transaction started by its parent container. Not Supported indicates that the container does not start a transaction or join an existing transaction.

64. What does the pivot transformation do?

It transforms rows into columns.

65. What is the difference between Fuzzy Grouping and Fuzzy Lookup?

Fuzzy grouping adds a column to the output of a table after examining the similarity between one or more columns and adds columns to the output describing the similarity index. The grouping will compare the rows to the first instance of the row that appears. A fuzzy lookup will perform a join/lookup based on whether the row's columns being compared are similar to a degree to the looked up values being joined on.

66. What is embedded SEQUEL?
It is SQL in your SSIS.

67. What are the types of transformations in SSIS?
Blocking – Will not return output row until it has received all the rows and processed them.
Non-Blocking – Will process each row and commit it individually. It does not use any of the buffer.
Partially-Blocking – It will not return any data until it reaches a particular level. Then the data will be processed.

68. What is a template package in SSIS?
A template package is a package that contains things that you want to reuse.

69. What command line tools can you use with SSIS?

DTutil, DTexec, and DTexecUI are three that can be used.

70. Types of containers in SSIS?

For Each container, For container, Sequence container, and Task host container.

71. What are common uses of containers?

Performing iterative processes or processing data sets.

72. How would you pass database information from SSIS to ASP.Net?

Using ASP.Net's datareader.

73. What file extension do SSIS files have?

dtsx

74. What is the prerequisite for data to be put in Merge and Merge Join transformations?

The data from the pipeline must be sorted.

75. What is SMSS used for?

It is used to deploy packets on the SQL server. It is what you use to interact with the database.

76. What is the difference between the For container, For Each container?

For Each will run through all the items in an enumeration. For container is used when you have bounds.

77. What is Visual Source Safe?

It is used for keeping backups of your project files, versioning them, and sharing them in between developers.

78. What is an ODS?

ODS is an abbreviation for Operational Data Store. It contains recent data that operations will be performed on. It normally contains close to 30 – 60 days of data. It does not contain static data.

79. What is the difference between a Fuzzy look up and look up?

The Lookup transformation uses an equi-join to locate matching records in the reference table. It returns either an exact match or nothing from the reference table. In contrast, the Fuzzy Lookup transformation uses fuzzy matching to return one or more close matches from the reference table.

80. What is the difference between a Merge and Merge Join operation?

A merge requires the two incoming data sets to have the same amount of columns. It then performs a union on the values. A merge join simply requires a common column on which it can perform a full join, left join, or inner join on.

81. What is the difference between using a dataflow task or bulk insert task?

Bulk insert doesn't do data validation, while the dataflow task will. You can't do transformations or error handling in bulk inserts, while you can data flows.

82. What is the difference between an EXECUTE PROCESS TASK and EXECUTE PACKAGE TASK?

In execute process tasks, you can execute batch files, and in execute package task you execute packages.

83. What is business intelligence?

Business intelligence (BI) mainly refers to computer-tools used in identifying, extracting, and analyzing business data, such as sales revenue by products and/or departments, or by associated costs and incomes.

84. Can you put an event handler on an event handler?

Yes you can.

85. What are the Slowly Changing Dimensions and what is supported and not supported?

Slowly changing dimensions are a way of tracking updates to the data warehouse records. Only type 1 and type 2 slowly changing dimensions are supported.

86. What does the derived column transformation do?

It applies a transformation to existing columns and puts it in a new column.

87. What is delay validation used for?

It is used when the value/or status of a task will be determined by a run time factor, and therefore verification of whether the task will work should be delayed to run time. This factor will be set on the package.

88. What are SSIS package template used for?

It is used when particular components in a package will be reused in different packages.

89. In what cases would you use a fuzzy lookup table?

When you have a reference table with clean values that you want to compare user input values against. This will compensate for error.

90. What are the two parameters in fuzzy lookup?

Similarity and Confidence.

91. What is the difference between a parameter and a variable?

A parameter allows you to assign values to properties within a package, while a variable holds values within it.

92. What are the different ways that SSIS offers to execute script tasks?

ActiveX Component, Script task, and Script component.

93. What are some ways to debug using SSIS?

a. Creating breakpoints on packages, containers, and tasks and suspend execution based on particular events.
b. Inspect the current state of execution, using debug windows such as Locals, Autos, and Watch.

c. Understand progress messages in Progress and Execution results tabs.

d. Partially execute tasks by disabling containers.

e. Set data viewers on data flow tasks to view data between transforms.

f. Create breakpoints on script elements and inspect code line my line using VSA.

94. How do you set breakpoints in SSIS?

Right click on the task object and click Edit Breakpoints. You will be presented with the possible breakpoints to set on the task.

95. How do you set checkpoints on a file?

By selecting the checkpoint's file location on the package level attributes.

96. What are some of the benefits of using checkpoints?

You will be able to start from the point of failure, rather than having to start the process from scratch. This will save time on resource heavy procedures.

97. What does using package password do?

It will prompt for a password before the package is run on the SQL server.

98. What is role based security on packages?

It prevents particular roles of users from having the ability to run a package.

99. Give examples of blocking transforms.

Aggregate and Sort.

100. Give some examples of partly blocking transforms.

Merge, Merge Join, Pivot, Unpivot, Lookup, Data mining Query and Union All. Note that in the case of partly blocking transforms, that the rows input will most likely not match the rows output.

101. What are the different scopes of variables in SSIS?

Package, container, task, or event handler.

102. Why is it not advisable to have all your variables as Package level?

The broader the scope, the more memory the variable will consume during run time.

103. Why are surrogate keys useful in data warehouses?

They are useful because when you obtain information from multiple databases, you might have redundant source keys. Therefore, you can keep track of the changes with the data warehouse having it's own keys.

104. What is the difference between an ETL tool and an ETL process?

The strategy is the plan, and the ETL tool is the vehicle that delivers the process.

105. What is the difference between an ETL tool and ETL package?

It is what is used to build a package. A package is a deliverable of an ETL tool. The package is an instance of my ETL strategy where the logic in implemented and delivered.

106. What is the two natures of every 'transformation'?

1. Blocking nature - B, NB, PB
2. Communication nature - How is communication conducted via upstream and downstream
   (determines whether it is synchronous or non-synchronous).

107. What is the most complex package you ever wrote?

Parent-child relationships. 50 tasks somewhere.

SSRS Questions

108. What is SSRS used for?
SQL Server Reporting Services (SSRS) is a report generating tool from Microsoft. It is used for generating reports.

109. How to optimize a SSRS report?
Optimize the query (joins instead of views), optimize sql query, paging, caching, or snapshots.

110. What is twinkling data?
It is when the data is updated in between reports being generated. Therefore, the reports that should be on the same data are not, and are slightly different. This normally occurs in the case of real time data warehouses.

111. What is the difference between a sub-report, linked report, drill through, drill down, and report?

A sub-report is a report within a report, a linked report is basically a report based on a perspective of another report (you set constraints on some parameters and generate a report from that), a drill through report is a report with bookmarks and links, a drill down report is a report that has toggles.

112. Where do SSRS snapshots vs cache reside?

Snapshots are permanent and stored in the ReportServer, and cache is temporary and stored in tempdb. Snapshots do not expire, the report will go to the snapshot even if the database changes.

113. What is the difference between caching and snapshots?

Caching is temporarily stored data of a report that is periodically refreshed. It is created based on the actions of a user and is used to speed up report performance. Snapshots are reports that are taken at a particular time. They are more permanent, and you can have more than one at a particular time, and you can use them to keep a history of how the report changed based on the database values over a period of time. It is not based on user actions, and you can have more than one at a given time.

114. How do you create an ad-hoc report?

You create a report model in Visual Studios, go to report server, go to report builder and create the ad hoc report on top of report model. The dataset of the ad hoc report must be in the report model.

115. What is the benefit of a parameterized report?

It allows you to choose to see what you want to see, rather than going through the entire report.

116. What databases does the report server use?

The databases are reportserver (stores the report catalogue) and reportservertempdb.

117. What types of roles are available in SSRS 2008, and what are their purposes?

Item-level roles and system-level roles are the two types of roles available in SSRS 2008. An item-level role is a collection of tasks related to operations on an object of the report object hierarchy of SSRS 2008. A system-level role is a collection of tasks related to operations on

server objects outside the report object
hierarchy of SSRS 2008.

118. How do you add drill down to reports?
Through using joins, variables and changing
the visibility on the groups of items which you
want to have drill down on.

119. How do you add drop downs when creating selective reports?
You add a variable reference in your SELECT
query for one of your datasets. You add
another dataset that will hold solely the values
of the drop down and get the values from a
query. Then you reference the value from a
parameter. This will automatically create a
parameterized report.

120. What is the expressions option under the dataset menu for a cell used?
Customizing how the value of a cell is
presented.

121. What is a data bar column?
It is a pictorial graphical representation of the
column value in comparison to the other
values in
the column.

122. In a report if we have data like 100.88899, how will you make sure that you round it off to two decimal places using SQL queries?

You can use the ROUND() function to round a column in a database.

123. What is a tablix?

Tablix is a flexible report item that can be used to display data in a grid format, with layout possibilities ranging from simple tables to advanced matrices. It is used to manage grid structures in SSIS.

124. What does selecting show data values on a data bar column do?

It will allow you to see the values next to the value bar in the data bar column rather than having a separate column specifically for it.

125. What is conditional formatting?

It is when you change the format of your report based on the value in your data set.

126. What are some considerations that you should make when designing your SSIS package?

Minimize data in the buffer since often, you will be working with a subset of what will be running when the system goes live. Remove redundant columns and data. Always use a SELECT statement rather than tables when configuring a data source. Remove columns if you are no longer using them.

127. Can the conditional split be used to concatenate columns?
Yes.

128. What is a drill through report?
It is when you use two different tables, and you use one table to navigate a different table.

129. How do you create a drill through report?
A drill through report is a report that a user opens by clicking a link within another report. Drill through reports commonly contain details about an item that is contained in an original summary report.

130. What is a SSRS snapshot?
A report snapshot provides the ability to run a report at a scheduled time or on demand and save the report in the ReportServer database. When a user requests the report, they automatically get the snapshot version stored in the ReportServer database rather than

executing the SQL queries to retrieve the data from the underlying data source. You can schedule report snapshots to be run during off-peak hours; this can reduce the load on your database servers.

131. What is a SSRS subscription?
An SSRS snapshot is a SSRS report that is generated on schedule in a selected format and delivered to the user in a particular fashion. It can put the report in a folder, or even email the report to them.

132. What is a bookmark and how do you create one?
A bookmark is used to uniquely identify a position in a report. In order to create a bookmark in SSRS, you would click on a column and in the properties explorer, you go to the bookmarks property. There you will set the expression you want to represent the bookmark.

133. How do you link data sets to a bookmark?
You would right click on the dataset, and set the following expression to the bookmark:
=Fields!ProductNumber.Value
or whatever you decided to label that particular bookmark. Next you would right click on the

dataset you wanted linked to the bookmark, and select text-box properties. In textbox properties, you would select Actions. There you would select the option, Go To Bookmark. Under select bookmark, you would select the bookmark you wanted to go to.

134. What is an adhoc report?
It is a report that is created on the fly.

135. What is a report model?
Is a abstract designed in SSRS to export to Report Manager, so that users can build Ad-Hoc reports on the fly.

SSAS Questions

136. What is a cube (Analysis Services Database)?
It is a database that lives on top of a data mart or data warehouse. It scans the underlying schema and extracts the data.

137. What is a virtual cube?
A virtual cube is a cube object that used to exist in older versions of Analysis Services. It was basically a view of a cube. It has been replaced by perspectives.

138. What is a data source view?

It will limit what tables can be altered and accessed by the system to the components in a view. It is the window to the source connection.

139. What are the benefits of an OLAP cube over a relational database?

Consistently fast response. Meta-data based queries. Spreadsheet based formulas.

140. What are spread sheet like OLAP applications usually associated with?

Financial applications.

141. What is a linked cube?

It is a cube inside of a cube. In order to create one you must have two cubes in your project, and you must go to Cube Structure, and add a linked object. In linked objects, you will select your other cube, and then add it. It will then become a linked cube.

142. What is the benefit of using a OLAP database model over OLAP spreadsheet model?

It avoids data explosion since the database structure is not solely 2D like spreadsheets.

143. What is data explosion?

It is a term to describe how OLAP systems are exponentially larger than the OLTP system they interact with. This is because the cube stores metadata, aggregations, and data.

144. What is MDX?

Multidimensional Expressions (MDX) is a query language for OLAP databases, much like SQL is a query language for relational databases. It is also a calculation language, with syntax similar to spreadsheet formulas.

145. What is ROLAP?

Relational Online Analytic Process.

146. When adding data from multiple sources, which source will become the primary?

SQL Server will always be the primary source.

147. What is MOLAP?

Multidimensional Online Analytic Process. It offers a speed increase over OLAP.

148. What is XMLA?

XML for Analysis (abbreviated as XMLA) is an industry standard for data access in analytical systems, such as OLAP and data mining.

149. What is DMX?
Data mining extensions. It is a language used for data mining.

150. What are the three types of facts?
Additive, Semi-additive, and non-additive. You can sum values over time (Savings). Semi-additive can only be summed at a snap shot of a time, not for all times (for example, the amount of money a group has each day). Non-additive measures can be ratios cannot be summed at any time at all. However, they can be averaged.

151. What is a SSAS Assembly?
They allow you to create external CLR user defined functions and add them to your cube. They are usually written in a .NET language.

152. What are perspectives?
It is similar to a view in SQL. When you want dimensions from the cube which shows particular things. It doesn't help with performance, however, it does help with visibility and navigation.

153. What are pros and cons of perspectives?

They make large cubes very user friendly. It is not security friendly, it will still allow you to connect to the entire cube. (Improve on this answer)

154. What are translations?

It is a method to translate table column results into another language.

155. What is a user defined hierarchy?

It is for navigation purposes only.

156. What is the difference between the key usage of regular, parent and key?

A parent child dimension is based on two dimension table columns that together define the lineage relationships among the members of the dimension. One column called the member key column identifies each member, the other column, called the parent key column identifies the parent of each member. If the key usage of a column is set to regular, it states the column is an attribute. To simplify, you can look at things as follows:
Parent = Foreign key (for the parent key)
Key = Primary key

regular = attributes

157. What is the difference between a data source view and a perspective?

Data source views are on the source connections. Cubes and dimensions are built on the data source view. The perspective is built on the cube.

158. What are the steps in creating a cube?

Go into BIDS, create a new Analysis Services Project.
Add the data source from which you want the cube to pull information.
Create a datasource view for the information to be accessible to the cube from the datasource.
Add the cube, and make it draw information from the DSV. Add the measures and the facts.
Add your aggregations, KPIs, hierarchies, calculated members, partitions etc … to the cube.

159. What tab do you create relations in with SSAS?

The Dimensions Usage tab.

160. What is a named query?

It is when you make a table in the data source view that is defined by a query (you right click and go to new Query)

161. What is KPI?
It is short for Key Performance Indicator. It is used in order to monitor whether milestones are met, trends are followed, and conditions are met by a particular company.

162. What are the types of hierarchy?
Natural hierarchy (one parent per child) and user defined hierarchy.

163. How do you localize your Analysis Services database?
Localization of a cube, can be done ahead using the underlying database or by using translations.

164. What are the three main storage modes?
The first default mode is MOLAP (Multidimensional OLAP). Best performance.
MOLAP = Data + Metadata + Aggregations
MOLAP stores aggregations only when they are designed.
The second mode is HOLAP (Hybrid OLAP).
HOLAP = Aggregations + Metadata

However, the data remains in the data source.

The third mode is ROLAP (Relational OLAP). Worst performance.

ROLAP = Metadata Only

The aggregations will be created in the underlying data source in the form of indexed views.

165. Bitmap index
It doesn't sort the data in a B-tree structure. Rather, it utilizes a grid format. At the moment these
do not exist in MS SQL.

166. When will aggregations be in a cube?
Aggregations will be present in a cube only when aggregations are designed.

167. Can MOLAP mode exist without aggregations? Can HOLAP exist without aggregations?
Yes

168. When is HOLAP and ROLAP similar?
When neither have aggregations they are similar.

169. Does ROLAP have processing?

No.

170. Where should you store a backup file in SQL Server?

Never store it on the same drive as the database. If that computer crashes, then you will not be able to access the backup.

171. Is ROLAP real-time?

No, because there is no processing, it is updated almost instantaneously. However, this is not true. A real time db is when a transaction is written in the Cube . It will take a bit of time doing this.

172. Do MOLAP and HOLAP process?

Yes

173. Why would a person use a HOLAP?

It is for massive amounts of data that are not used often.

174. Why would a person use ROLAP?

This is for data that is used even less than HOLAP. This is for very old data.

175. What are aggregations?

It is a collection of tables that holds values with members of the same granularity.

Types: Default, Full, None and
unrestricted
Id
First name
Last name
Sum of id etc …
Sum of fn etc ..
Sum of ln etc ..
Do not create aggregation
Do not create aggregation
Do not create aggregation
Sum of title etc ..
title
Do not create aggregation
Default – if aggregations already exists SSAS
will not create the same aggregations again.
Mainly default and unrestricted allows analysis
services to figure out things.

176. What does setting the aggregations performance value do?
Creates more aggregations in order to speed
up the queries for different situations.

177. What is the difference between processing and deploying?
Processing is loading the values, and
deploying is loading the structures.

178. How do you make your cube interactive?

You can do this by creating actions on the cube.

179. What is the relation between fact tables and partitions?

Each measure group (fact table) has it's own partition in SASS.

180. What is a query bound partition?

It is a partition that returns only a particular set of records specified by the user. However, you cannot have overlapping or gapping partitions.

181. For you to use partitioning in SQL or Microsoft analysis services, what version do you need?

You will need the Microsoft Enterprise Edition. The developer edition cannot support partitioning.

182. When would you use a degenerated dimension?

When you have a dimension that grows at the same rate as the fact table. Then you combine them since they have the same granularity rather than having two tables. Thus, a degenerated dimension doesn't have it's own dimension table. It's attributes are in other dimensions. In a degenerated dimension, you have a primary key and all it's attributes are foreign keys. Thus, since the attributes are already referenced in other places, you can access them in an indirect manner. Also, you take the primary key and bring it in to the fact table since you no longer need the foreign key that points to the table.

183. What is a degenerated dimension?
A degenerated dimension is a design technique that creates a dimension inside a fact table. Where the dimension and the fact share the same granularity and the attributes of that dimension is distributed across dimensions connected to the fact table.

184. What is a factless fact?
It is a fact table with no measures. It is similar to a conjunction table in OLTP.

185. How in analysis services can you localize your cube?
You can use translations in order to return the cube results in the language of the place you are in. Translations can be accessed via the translation tab in Analytics.

186. What are the pros and cons of having perspectives and how would you use them?

Pros

A perspective is a definition that allows users to see a cube in a simpler way. A perspective is a subset of the features of a cube. A perspective enables administrators to create views of a cube, helping users to focus on the most relevant data for them. A perspective contains subsets of all objects from a cube.

Cons

A perspective cannot include elements that are not defined in the parent cube. You cannot have additional security on a perspective. If you have access to the perspective, you have access to the entire cube. It is read only.

187. What is the difference between full and default aggregations?

A full aggregation creates all the aggregations for each and every one of the tables selected. A default aggregation will not create the aggregation if the aggregation exists in another place.

188. What are the benefits of partitioning a cube measure group/partition?

The processing time required for large measure groups can be reduced when we partition those groups, because processing can then be undertaken in parallel across the partitions. (Parallel processing means faster execution, primarily because the processing of one partition does not have to finish before the processing of another can start; more than one processing job can run at the same time, typically utilizing processer capacity more efficiently.) And when we distribute the data over multiple machines with remote partitions, we not only provide more physical room for large volumes of data, but we make it possible for multiple computers to process the data in parallel.

189. What is the difference between a table bound and query bound partition?

A table bound partition is a partition that is bound to a fact table. A query bound partition is when the contents of a partition is defined by a query (when doing this you must make sure that you do not overlook any of the data).

190. How do you format your measures in a measure group? What is the name of the property and under which tab does it fall under?

You can format your measure using the Cube Structures tab in analytics. You can select the measure and edit the 'Format' property in the properties window.

191. Under the cube object, which tab would you use to specify the relationships between dimensions and fact tables/measure groups?
Dimension Usage.

192. What is an additive measure?
It is a measure that can be added over time. Profits and costs.

193. What is a semi-additive measure?
It is added at a snapshot of time. For example, rooms unoccupied.

194. What is a non-additive measure?
It is a ratio of rooms unoccupied. These are usually ratios.

195. What is the difference between referenced, many to many, fact, and regular relationships in analysis services?
A regular dimension relationship between a cube dimension and a measure group exists when the key column for the dimension is

joined directly to the fact table. This direct relationship is based on a primary key–foreign key relationship in the underlying relational database, but might also be based on a logical relationship that is defined in the data source view.

Fact dimensions relationships, frequently referred to as degenerate dimensions, are standard dimensions that are constructed from attribute columns in fact tables instead of from attribute columns in dimension tables. Useful dimensional data is sometimes stored in a fact table to reduce duplication. A single fact table joined to multiple dimension members is called a many-to-many relationship. A reference dimension relationship between a cube dimension and a measure group exists when the key column for the dimension is joined indirectly to the fact table through a key in another dimension table.

196. What are actions in SSAS?
Actions enable business users to act upon the outcomes of their analyses. By saving and reusing actions, end users can go beyond traditional analysis, which typically ends with presentation of data, and initiate solutions to discovered problems and deficiencies, thereby

extending the business intelligence application beyond the cube. Actions can transform the client application from a sophisticated data rendering tool into an integral part of the enterprise's operational system. Instead of focusing on sending data as input to operational applications, end users can "close the loop" on the decision-making process. This ability to transform analytical data into decisions is crucial to the successful business intelligence application.

You can exercise flexibility when you create actions: for example, an action can launch an application, or retrieve information from a database. You can configure an action to be triggered from almost any part of a cube, including dimensions, levels, members, and cells, or create multiple actions for the same portion of a cube. You can also pass string parameters to the launched applications and specify the captions displayed to end users as the action runs.

197. How do you make your cube interactive?
You can do this by creating actions on the cube.

198. What is a degenerated dimension?

A degenerated dimension is a design technique that creates a dimension inside a fact table. Where the dimension and the fact share the same granularity and the attributes of that dimension is distributed across dimensions connected to the fact table.

199. What storage model tracks the business processes?
OLTP

200. What storage model analysis the business processes?
OLAP

201. What do KPIs allow?
It allows business analysts to make intelligent decisions based on KPI gauges placed on business processes in your database.

202. When adding semi-additive measures, what is the requirement in analysis services to add or change the aggregated function of a measure to a semi-additive function?
One dimension has to be flagged as a time dimension. Which means we need to add time business intelligence to the time dimension.

203. What does lastnotempty mean?

It means the current value as of the moment.

204. How do you know if you have dirty data?

You need to have a subject matter expert tell you since you cannot tell if data is right or wrong.

205. What is clean data?

It is data that has correct values, relationships and metadata.

206. What is a namecolumn for a property in SSAS?

A name column is column that is refered to by another column. It is used by the CUBE browser to substitute the value in the name column in place of the value of the actual column. This is used to make CUBE browsing easier.

207. What is a keycolumn for a property in SSAS?

A key column is used to define which attributes are related.

208. Why do attribute relationships require hierarchies?

It helps to speed up querying of the cube, and tells the system which attributes are related and how they are related. This will tell the CUBE how to allow users to navigate the CUBE's information.

209. How do you handle a very large Analysis Services Database?

Two ways to optimize. The reading, and the processing. You can analyse quantity of data, cube size of quantity of data, the underlying and existing infrastructure and the various external parts. Based on this you can make a strategy and utilize ROLAP, HOLAP, MOLAP (Processing).

210. What is adding business intelligence to a dimension?

This will tell the cube what type of dimension it is and help the cube better analyze the data.

211. What are the different storage modes of dimensions?

MOLAP, ROLAP and InMemory. Dimensions do not have HOLAP.

212. Different ways of identifying relationships between fact tables and dimensions?

In the underlying database, the data source view, or the dimension usage tab.

213. How do you define a relationship between a degenerated dimension and a fact table (a fact relationship)?

You would reference the dimension to fact in the dimension usage tab.

214. What is a named query?

Is a user defined SQL query that can be used as its own table in the data source view. You can derive a desired dataset or even change the structure of the connections (change a snow flake schema to a star schema). Remember that the cube will only store the result set, not the query.

215. How do you identify a time dimension?

You can either add business intelligence (this will also allow you to map existing attributes to the wizards attribute) or change the type of the dimension in it's properties.

216. How do you create or go about creating a parent child dimension?

Define it in the data source view, or in the under lying database, or in the dimension pane. Remember it is not in dimension usage (it has to be between dimension and fact here).

217. If you create a calculated member, however, it is not viewable when you are browsing the cube. What could be the reasons?
It might be that you need to reconnect to the source, the visibility may not be true or false, you may have had a perspective view that does not show the cube, or you haven't deployed.

218. True or False. Measures in query bound partitions can overlap.
False.

219. True or False. Measures in query bound partitions can have gaps.
False.

220. What is data profiling, data cleansing, and blind loading.
Data Profiling – Trying to understand the most you can understand about the metadata of data (stored procedures, DMVs, etc ... finding nulls, ...). All data on the metadata of your sources. [ETL map data profiling]

221. Problems with extraction of data?

Data type errors, conversion errors, Oracle (no permissions to extract data, network problems, may loose packets). Flat files – delimiter separation (different from what you are used to). SQL server (same problems).

222. What is the most complex stored procedure, package, cube, etc ... you have ever created?

Stored procedure (1000 lines of code), challenging to debug it, different error handling and different tables, nested stored procedures, data management views, role operators, set operators, indexes, temp tables, SQL profiler [DTA] – suggests indexes, used for data profiling.

223. Are actions accessible from any cube browser?

Yes they are.

224. What are thin clients?

They are third party tools that accesses the cube in order to read and present it's data.

225. What is a way to process a cube?

You can right click on the cube and process, command line, you can use the task in the SSIS toolbox and make it a job.

226. What are the different expressions in a KPI?

Value expression, goal expression, status expression and trend expression.

227. Give examples of structure changes in SASS.

A formula: Process Full = Process Structure, Process Data, Process Indexes
Example:
New perspective
Change edit KPI = structure change
New relationship between dimension and cube
Add or delete dimension
Any change in metadata = structure change

228. What partition methodology is specific to cubes?

Partition Increment.

229. What is the difference between process update and process incremental?

- process update is used for data in dimensions. when change attribute

- process incremental used for data in measures in cube and process update deals with structure also.

230. What are the prerequisites of process index?

All the objects the index is placed on should be processed ahead of time.

231. What are the different ways to process?

Process index, process data, unprocess, process structure, process clear structure, process script cache, process update, process clear structure, process script cache and process incremental.

232. Scenario: If you were manually scheduling the processing of a cube. The size of the cube is 850 gigs. The underlying data warehouse gets updated every 20 minutes with an increment averaging 20 megabytes of data. The cube has 9 years of historical data. Two measure groups, 8 dimensions, the cube is supposed to reflect the data immediately after it has been loaded to the data warehouse. Create a processing strategy.

Now days SSIS Jobs are used to update the cube via incremental updates, while in the past you'd set the updates within the cube settings itself. You will separate the storage of the cube according to the storage modes (ROLAP, HOLAP, and MOLAP).
(You should improve on this answer)

234. What is the scope of a cursor?
It is session bound.

233. How much faster is a cube than the underlying data warehouse?
17 – 32 times faster.

SQL Profiler

1. What is SQL Profiler used for?
Finding database bottlenecks, debugging stored procedures and user defined processes, keeping a log of the database activity, identifying deadlocks, identifying blocking and much much more ... It is a database tool used by developers, QA testers and DBAs.

2. Must SQL Profiler be run via GUI or Is there another way to run SQL Profiler?

You can run SQL Profiler via T-SQL commands. Therefore, you can make Profiler it run in the background.

3. What is a template in SQL Profiler?
It is a predefined configuration for SQL Profiler that will make it monitor specific database activity.

4. What a filter in SQL Profiler?
It is a condition on the SQL Profiler data that will control what data Profiler records.

Real Interview Questions

Since these were interview questions, many will not have answers associated with them. Feel free to research them and fill in the blanks.

1. What would your colleagues describe you?
2. Where do you see yourself five years from now?
3. Why are you seeking this position?
4. When did you complete your last project?

5. Name the most popular ERP software suite in use today.

6. How big were the databases you worked with?

7. What is SPRINT and SCRUM?

8. Have you used agile methodology?

9. Give me an example of how you went about testing your ETL Package.

10. Did you work directly with real data?

11. How can you make sure that your underlying data source is protected i.e prevent sql injection?

12. How did you do data profiling?

13. What is a fact table vs a dimension table?

14. What is the difference between stress ad load testing?

15. How did you write a test plan?

16. What does test plan contain and what is it?

17. How do you decide how much time a test is going to take?

18. What are the tests you have performed?

19. How will you decide if a test is complete or not?

20. Talk about priority, severity, suspension criterion, resumption criterion.

21. What is performance testing?

22. How to trouble shoot slow performance of applications?
23. What is Agile Development?
24. What are Use Cases?
25. How big was your team, responsibility, and title?

Team Size – 12
Project Manager - 1
Team Lead – 1
Business Analyst - 1
DW Architect - 1
Data Analysts/Data Mapper – 2
SQL Developer - 2
BI Developer including ME - 3
Tester/QA -1

30. How big is your data warehouse?
About 500 GB, if a big company it may be as large as 20 Terabytes (Bank of America)

31. At what rate did it grow?
For bank of America it can be as high as 13 GBs a day.

32. What are some stored procedures you created?
33. What is a fail over cluster?

34. Describe the most complex stored procedure you've created and why it is most complex?

35. Describe the environment.

SQL Server 2000/2005/2008, Window 2003 Server, SSIS, SSAS, SSRS, BI Development Studio (BIDS), VS.

36. Describe some reports you've created?

37. Describe the most complex ETL package you've created.

38. In your cube, what were some examples of KPIs you used?

I am not allowed to talk about this.

39. What are the types of hierarchy you've created?

40. Do you have any confirmed dimension?

DimTime

41. Give examples of role playing dimensions you worked with.

DimTime, I am hesitant to talk about the design in more detail about any of the othe dimensions.

42. What was the granularity level of each fact table?

43. What is SQL powershell?

An interface that allows you to browse through the network like command prompt.

44. Why do you want to work for us?
45. Who usually uses the OLTP and OLAP?

OLTP is for end-consumer, and OLAP is for managers and business analysts.

46. What is a key attribute?

They are attributes that the other attributes are dependent on in a table.

47. What is a CRM?
48. What is an ERP?